What if you Could

SPY LIKE A NARWHAL!?

EXPLORE THE SUPERPOWERS OF AMAZING ANIMALS

by
Sandra Markle

illustrated by
Howard McWilliam

Scholastic Inc.

**For Joe Merrill and all the children
of Lake Park Elementary in Naples, Florida.**

*The author would like to thank the following people for sharing their expertise and enthusiasm: Dr. Nina Attias, Federal University of Mato Grosso do Sul, Mato Grosso do Sul, Brazil (Three-banded Armadillo); Dr. David Blackburn, Florida Museum of Natural History, Gainesville, Florida (Hairy Frog); Dr. Anastasia Dalziell, University of Wollongong, Wollongong, NSW, Australia (Superb Lyrebird); Dr. Gerhard von der Emde, University of Bonn, Bonn, Germany (Electric Eel); Dr. Patrick Flammang, Université de Mons, Mons, Belgium (Sea Cucumber); Dr. Stefano Grignolio, University of Sassari, Sassari, Italy (Alpine Ibex); Dr. Jens C. Koblitz, BioAcoustics Network, Neuss, Germany (Narwhal); Dr. Kristin Laidre, University of Washington, Seattle, Washington (Narwhal); Dr. Dzulehlmi Muhammad Nasir, University of Malaya, Kuala Lumpur, Malaysia (Colugo); Dr. Alexandra Schnell, Cambridge University, London, United Kingdom (Cuttlefish).
A special thank you to Skip Jeffery for his support during the creative process.*

Photos ©: cover, 1 wood sign: Dim Dimich/Shutterstock; cover background: kzww/Shutterstock; cover background: H. Mark Weidman Photography/Alamy Stock Photo; 4: Todd Mintz/Alamy Stock Photo; 6: Animals Animals/Superstock, Inc.; 6-7: dotted zebra/Alamy Stock Photo; 7: Bryan & Cherry Alexander/BluePlanetArchive.com; 8: Julian Kaesler/Getty Images; 10: D. Parer and E. Parer-Cook/Minden Pictures; 10-11: Arco Images GmbH/Alamy Stock Photo; 11: tracielouise/Getty Images; 12: Bonnie Fink/Shutterstock; 14: Belizar/Dreamstime; 14-15: Eric Isselee/Shutterstock; 15: Robert Eastman/Alamy Stock Photo; 16: Alessandro Oggioni/Shutterstock; 18: Janko Bartolec/Dreamstime; 18-19: Pixtal/Superstock, Inc.; 19: Andrea and Antonella Ferrari/NHPA/Avalon.red/Newscom; 20: David Yeo/Getty Images; 22: Puntasit Choksawatdikorn/Dreamstime; 22-23: Oliver Thompson-Holmes/Alamy Stock Photo; 23: Joshua Davenport/Shutterstock; 24: Mark Newman/Getty Images; 26-27: Paulo Oliveira/Alamy Stock Photo; 27: Amazon-Images MBSI/Alamy Stock Photo; 28: Paul Starosta/Getty Images; 30: Paul Starosta/Getty Images; 30-31: Paul Starosta/Getty Images; 31: Paul Starosta/Getty Images; 32: Genevieve Vallee/Alamy Stock Photo; 34: Fred Bavendam/Minden Pictures; 34-35: AF archive/Alamy Stock Photo; 35: Irko Van Der Heide/Dreamstime; 36: cbimages/Alamy Stock Photo; 38: D.P. Wilson/FLPA/Science Source; 38-39: Georgette Douwma/Minden Pictures; 39: Mark Conlin/Alamy Stock Photo.

Text copyright © 2021 by Sandra Markle
Illustrations copyright © 2021 by Howard McWilliam

Library of Congress Cataloging-in-Publication Data available

ISBN 978-1-338-35609-0 (paperback) / 978-1-338-35610-6 (hardcover)

10 9 8 7 6 5 4 3 2 1   21 22 23 24 25

Printed in Malaysia   108
First edition, January 2021

Book design by Steve Ponzo.
Photo Research by Marybeth Kavanagh.

What if one day when you woke up, you found out that overnight you gained a WEIRD animal SUPERPOWER? What if you could generate powerful electric shocks like an electric eel? Vanish like a giant cuttlefish? Or mimic all kinds of strange sounds like a superb lyrebird? How in the world could a weird animal superpower change YOUR life!?

WHAT IF YOU COULD SPY LIKE A NARWHAL?

WHERE IN THE WORLD?

Narwhals live in the Arctic Ocean.

A narwhal doesn't spy with its eyes the way you do. A special body part just below its blowhole produces streams of clicks, which pass out through its forehead as sound beams. A narwhal aims these beams and listens for echoes, which it can detect from nearly a mile away. This kind of spying is called echolocation. A narwhal uses echolocation to sense what's around it, find food, and survive. That's because, though a narwhal can hold its breath, it must surface for air. So, in winter when the Arctic Ocean is covered with thick ice, a narwhal uses echolocation from the dark depths to home in on a crack. Then, in the nick of time, it swims up, pokes its head out of the water, and breathes.

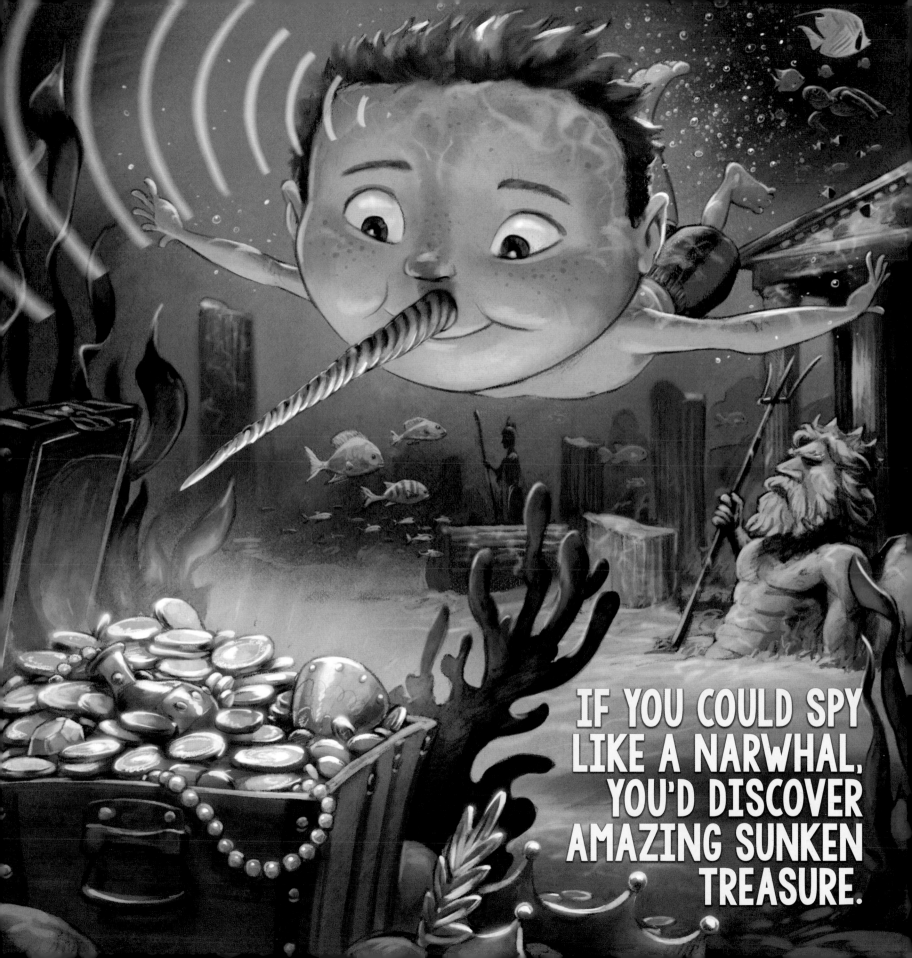

IF YOU COULD SPY LIKE A NARWHAL, YOU'D DISCOVER AMAZING SUNKEN TREASURE.

Adult Size
Up to 17 feet long and weighs up to 2,000 pounds

Life Span
Up to 90 years

Diet
Mainly fish, squid, and shrimp

fluke
Moves up and down to propel the narwhal through the water.

flipper

GROWING UP

A baby narwhal is called a calf. Females usually give birth to a single calf every three years. The calf develops inside its mother's body for about 14 months, and most are born in the Arctic spring (April through May). The young stay close to their mothers and nurse for at least a year as they learn to hunt on their own within the pod, or group of narwhals. Narwhals change color as they age. Calves are mainly gray, juveniles are completely bluish-black, mature adults are spotty gray, and older narwhals are almost completely white.

WONDER WHY?
a male narwhal has a long tusk?

What looks like a tusk is really its left canine tooth that's grown through its upper lip. It may grow to be up to ten feet long. Scientists believe males use their tusks to compete for mates. However, they have been observed dueling only with gentle tusk rubs. After all, hard knocks could cause a tusk-ache. OUCH!

blowhole

tusk
Usually found on males; only
rarely on females.

SUPERCHARGED!

If you could spy like a narwhal, you'd always find the shortest way out of the maze.

WHAT IF YOU COULD MIMIC SOUNDS LIKE A SUPERB LYREBIRD?

WHERE IN THE WORLD?

Superb lyrebirds live in wet forests in Australia.

Superb lyrebirds are talented sound mimics. Both males and females can mimic, but males do so more often—and they are louder! Male superb lyrebirds have been recorded mimicking the wingbeats of flying birds, chattering forest animals, and a chorus of singing kookaburras. Male superb lyrebirds in zoos have been recorded mimicking even more unusual sounds, such as a chainsaw, a car alarm, and a pounding hammer. Though scientists know the lyrebird's syrinx (voice box) is different than that of other birds, they're still studying how they are able to produce such unusual sounds. They do know that a female lyrebird follows the sounds that wow her to choose her mate.

WHAT YOU SHOULD KNOW

Adult Size
Males are about 32 to 39 inches long from beak to tail tip and weigh about two pounds. Females are smaller, with a plain, slightly shorter tail.

Life Span
Up to 30 years

Diet
Mainly worms and insects dug from the forest floor

beak

wing

leg

foot

GROWING UP

A baby lyrebird is called a chick. After mating, the female lyrebird leaves the male and lays a single egg in a dome-shaped nest of sticks she builds on the forest floor, sometimes on a tree stump or on a rocky ledge. Every day, she has to leave the nest to eat because it takes the chick about 50 days to hatch. After the chick hatches, the female brings food to the chick for six to ten more weeks. She also guards it from hungry hunters—predator birds such as pied currawongs—by mimicking alarm calls of other animals. Even after the young bird develops feathers and begins finding food for itself, it stays with its mother for most of its first year.

tail

Males first develop their amazing tails when they are three to four years old. Then they shed and regrow new tail feathers each year just in time for the mating season— June and July.

WONDER WHY?

a male superb lyrebird has such a showy tail?

A male superb lyrebird needs his showy tail to win a mate. Once his songs and sound mimicry draw a female close, he works hard to impress her by spreading his tail— even flipping it over his head.

SUPERCHARGED!

If you could mimic sounds like a superb lyrebird, you wouldn't need an instrument to play in the band.

WHAT IF YOU COULD ARMOR UP LIKE A THREE-BANDED ARMADILLO?

WHERE IN THE WORLD?

Three-banded armadillos live in South America, mainly in savannas and dry forests.

All armadillos have armor-like plates (together called a carapace) covering their bodies. But only the three-banded armadillo can ball up completely. It can do so because its head, back, and tail carapace plates are perfectly shaped to fit together. Strong muscles along its sides power this defensive move. When it first senses danger, usually it only rolls up halfway. That allows its nose to stay untucked for easier breathing. But if a big predator, such as a jaguar, comes close, its muscles jerk. SNAP! The three-banded armadillo's body rolls into a tight, armored ball that's too big and hard for the predator to bite.

IF YOU COULD ARMOR UP LIKE A THREE-BANDED ARMADILLO, YOU'D HAVE ALL YOU NEED TO DO COOL TRICKS AT THE SKATEBOARD PARK.

WHAT YOU SHOULD KNOW

Adult Size
Males are a little bigger than females; they may be as long as 15 inches from snout to tail tip and weigh a little over three pounds.

Life Span
Up to 20 years in zoos, but unknown in the wild

Diet
Mainly ants and termites.

hairy unarmored belly

tail

GROWING UP

A baby armadillo is called a pup. Three-banded armadillo mothers produce one golf ball–sized pup at a time. Scientists believe being pregnant with more would make it hard for the mother to ball up. The pup develops inside its mother's body for about 120 days. When it's born, it looks like a little adult, although its carapace is leathery soft for the first few days. The mother cares for the pup alone, protecting it and nursing it. During most of the rest of its first year, the mother and juvenile often share the same shelter at night.

carapace

WONDER WHY?
the three-banded armadillo has super strong, long claws?

Where it lives the ground is often stone-hard, but its strong claws let it dig almost two feet deep for an ant or termite dinner. Its claws are also so long its front feet walk on its claw tips rather than its footpad. Scientists call this a ballerina walk.

mouth
Its tongue is long, pointed, and coated with sticky saliva—perfect for grabbing and holding onto ants and termites.

claws

SUPERCHARGED!

If you could armor up like a three-banded armadillo, you'd be a star goal tender for the hockey team.

WHAT IF YOU COULD CLIMB LIKE AN ALPINE IBEX?

WHERE IN THE WORLD?

Alpine ibex live in the Alps on rocky slopes and cliffs and in mountain meadows.

An Alpine ibex is a climbing machine! Its short legs and body shape let it easily balance on narrow ledges. Alpine ibex are also strong and nimble enough to jump more than six feet from a standing start across a section of missing ledge. Most important, it has mountain-climbing feet—hooves made up of two toes that can move independently. Each toe section of the hoof has a hard, sharp outer edge for gripping, plus a soft middle part that acts like a nonslip suction cup.

IF YOU COULD CLIMB LIKE AN ALPINE IBEX, YOU'D CLIMB FUN WALLS AND HAVE YOUR SNACKS, TOO.

WHAT YOU SHOULD KNOW

Adult Size
Males are about twice as big as females. A large male's body is about five or six feet long and weighs around 176 pounds.

Life Span
About 14 to 18 years

Diet
Mainly grass

head

horn

beard
(on males)

hoof

GROWING UP

A baby Alpine ibex is called a kid. A female gives birth to one or two kids at a time after they develop inside her body for about 5½ months. Almost immediately after a kid is born, it is alert and able to run and jump and follow its mother, even on narrow mountain ledges. The kids nurse for about four to six months, although they are already sampling grasses and their horns are beginning to grow. When they're two to three years old, males join a nearby male herd. Young females stay with their mother's herd. The two herds usually meet only during the breeding season, December through January.

WONDER WHY?

both male and female Alpine ibex have horns?

Their horns help the adults stand their ground against other adults. Even more important, their horns help them protect themselves and their young from predators such as wolves. A male's horns can grow to be about 40 inches long; a female's horns usually don't grow longer than 14 inches.

body

Coat length varies with the seasons: In summer it's short and fuzzy; in winter it's thicker with long guard hairs. Females are light brown year-round; males are gray in summer and brown in winter.

SUPERCHARGED!

If you could climb like an Alpine ibex, you could easily retrieve a ball from the roof.

WHAT IF YOU COULD FLY LIKE A COLUGO?

WHERE IN THE WORLD?

Colugos live in forests in Southeast Asia.

When a colugo stretches out, it unfolds a giant skin sheet, called a patagium (puh-TAY-gee-um). This extends among all four legs and the colugo's tail. Even its toes are webbed to add air-catching surface. Plus, a colugo is very lightweight. No wonder that, in one leap, it can glide through the forest farther than the length of an American football field. WHOOSH!

IF YOU COULD FLY LIKE A COLUGO, TAKING OUT THE TRASH WOULD BE AN ADVENTURE.

Adult Size
Females are a little bigger than males—up to 16 inches long—but only weigh about 35 ounces

Life Span
Up to 15 years

Diet
Mainly leaves, flowers, and the tender new shoots at the ends of branches

eye ——

GROWING UP

A baby colugo is just called a baby. Colugos are believed to mate throughout the year. A mother usually gives birth to a single baby after the youngster develops inside her body for about 60 days. When resting, the mother curls up her tail, folding her patagium into a pouch for her baby. This keeps the youngster warm and safe from predators, such as owls or eagles. When mother travels, baby rides along by tightly gripping her furry chest. Though increasingly on its own, a young colugo takes about three years to mature.

patagium

leg

tail
Wiggling its tail slows down its glide speed so it can ease into a landing.

WONDER WHY?
a colugo has big eyes?

A colugo is mainly active at night. Having big eyes helps it find food and watch out for predators.

SUPERCHARGED!

If you could fly like a colugo, you'd be first in line every time.

ICE

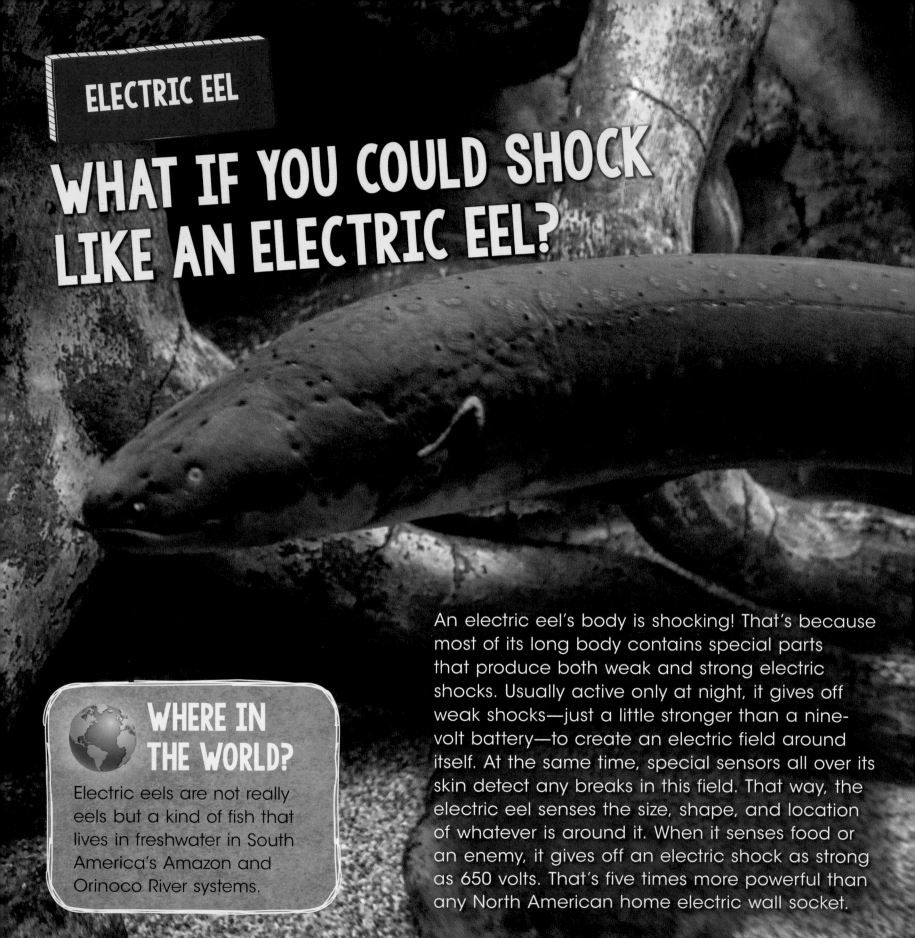

ELECTRIC EEL

WHAT IF YOU COULD SHOCK LIKE AN ELECTRIC EEL?

🌐 WHERE IN THE WORLD?

Electric eels are not really eels but a kind of fish that lives in freshwater in South America's Amazon and Orinoco River systems.

An electric eel's body is shocking! That's because most of its long body contains special parts that produce both weak and strong electric shocks. Usually active only at night, it gives off weak shocks—just a little stronger than a nine-volt battery—to create an electric field around itself. At the same time, special sensors all over its skin detect any breaks in this field. That way, the electric eel senses the size, shape, and location of whatever is around it. When it senses food or an enemy, it gives off an electric shock as strong as 650 volts. That's five times more powerful than any North American home electric wall socket.

IF YOU COULD SHOCK LIKE AN ELECTRIC EEL, YOU'D NEVER WORRY ABOUT THE POWER GOING OUT.

WHAT YOU SHOULD KNOW

Adult Size
Adults may be as long as 8 feet and weigh as much as 44 pounds.

Life Span
Up to 20 years

Diet
Mainly fish

head with all main body organs

skin

mouth

eye

GROWING UP

A baby electric eel is called a larva. Larvae develop inside eggs. The electric eel father creates a foamy nest by producing a lot of spit bubbles. Inside this nest, the mother deposits as many as 1,700 eggs, and the father covers them with sperm. Although it starts out as bubbles, this nest becomes tough so it doesn't break up as it floats on the river. The father guards the nest. About two weeks later, the larvae hatch. Though tiny, they can already produce a weak electric shock to catch equally tiny prey, such as some crustaceans. When predators such as bigger fish are around, the young electric eels go back inside the nest, still guarded by their father. He keeps guarding the nest and his young until the rainy season arrives about two months later. By then the young electric eels are big enough to leave and take care of themselves.

?

No photo currently exists.

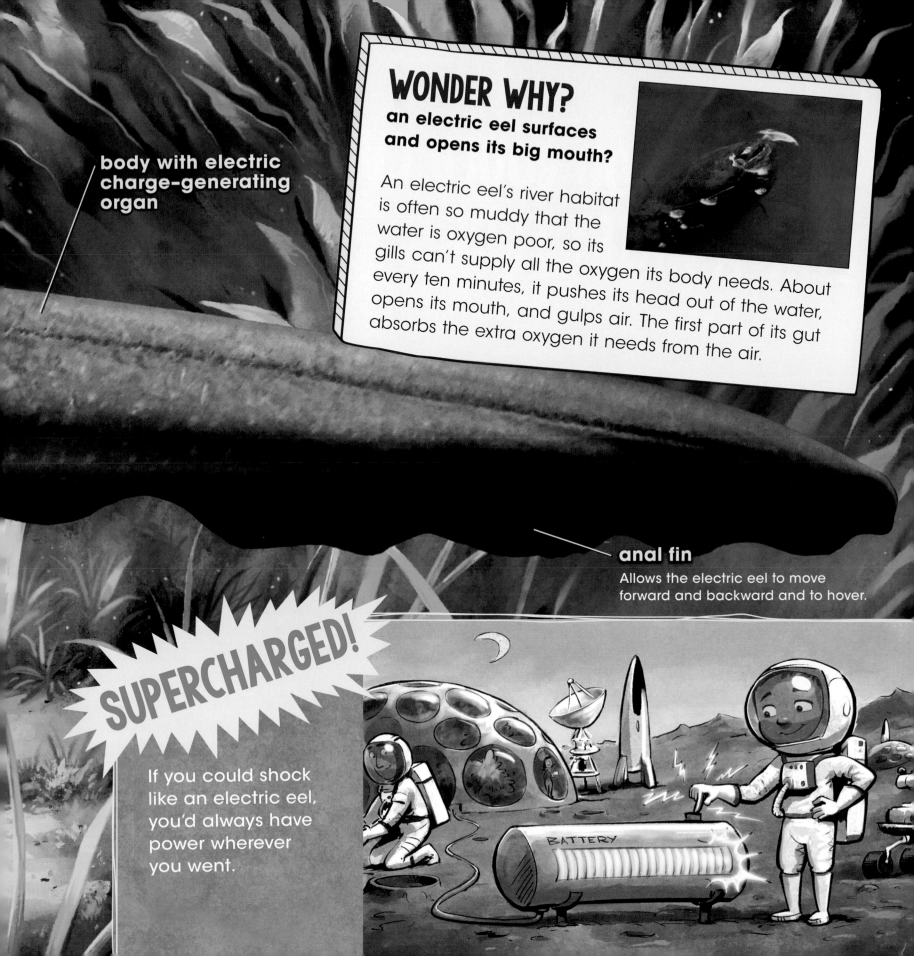

body with electric charge-generating organ

WONDER WHY?
an electric eel surfaces and opens its big mouth?

An electric eel's river habitat is often so muddy that the water is oxygen poor, so its gills can't supply all the oxygen its body needs. About every ten minutes, it pushes its head out of the water, opens its mouth, and gulps air. The first part of its gut absorbs the extra oxygen it needs from the air.

anal fin
Allows the electric eel to move forward and backward and to hover.

SUPERCHARGED!

If you could shock like an electric eel, you'd always have power wherever you went.

BATTERY

WHAT IF YOU COULD CUT LIKE A HAIRY FROG?

WHERE IN THE WORLD?

Hairy frogs live in tropical rainforests in central Africa.

Nobody messes with a hairy frog! Both males and females are armed with catlike extendable claws tucked inside the skin of the toes of their back feet. When a hairy frog is attacked by a hungry animal, such as an African otter, muscles attached to the stiff tips of each of its toes contract. That allows the bones at the toe tips to drive straight through the toe pads. Those bones are claw-sharp and they stab the frog's attacker. This surprising attack is usually enough to drive away its enemy. Scientists wonder—but don't yet know—whether a hairy frog's toes heal over or the claws remain out forever. Either way, a hairy frog is always ready to defend itself.

IF YOU COULD CUT LIKE A HAIRY FROG, YOU'D BE A FAMOUS SALAD CHEF.

eye

tympanum
(eardrum)

front foot

hair-like strands

Adult Size
About four inches long and weigh about 3.5 ounces. Males are a little bigger than females.

Life Span
Up to five years

Diet
Mainly small insects, millipedes, and snails

GROWING UP

A baby hairy frog is called a tadpole. A mating female and male release eggs and sperm in quiet waters along the edges of fast-flowing streams. Once the tadpoles hatch, they breathe through gills and swim into the quickly moving water. There, each uses its large mouth like a suction cup to anchor onto a rock. This allows the tadpoles to stay safe as hungry fish that might eat them are swept past. However, a hairy frog tadpole has to let go and allow itself to be swept along whenever it's catching its own meals of tiny insects and smaller tadpoles. Scientists aren't sure how long it takes a hairy frog tadpole to become a froglet. Froglets have developed lungs and they need to breathe air. Females leave the stream, but males remain along the stream's edge.

WONDER WHY?

a male hairy frog is hairy?

Only males have what looks like hair on their sides and back legs. These are actually thin strands of skin. Males appear to become even hairier during mating season. Scientists believe one reason male hairy frogs are "hairy" may be to attract a mate. However, scientists are still studying these frogs and have a lot left to learn!

hind foot

SUPERCHARGED!

If you could cut like a hairy frog, you could start your own pet grooming business.

GIANT CUTTLEFISH

WHAT IF YOU COULD VANISH LIKE A GIANT CUTTLEFISH?

WHERE IN THE WORLD?

Giant cuttlefish live in temperate ocean water, mainly along the southern coast of Australia.

A giant cuttlefish can appear to vanish because it's a hide-and-seek master. Its skin contains small sacs of pigment, or coloring material. These sacs are surrounded by muscles that connect to its nervous system. The muscles make the sacs expand or shrink to alter the cuttlefish's coloring so it closely matches its surroundings—even if they are not plain. And, because a giant cuttlefish has lots of muscles but lacks a hard, bony skeleton, it quickly changes its body shape to be the best possible match to its hiding place, helping it disappear. POOF!

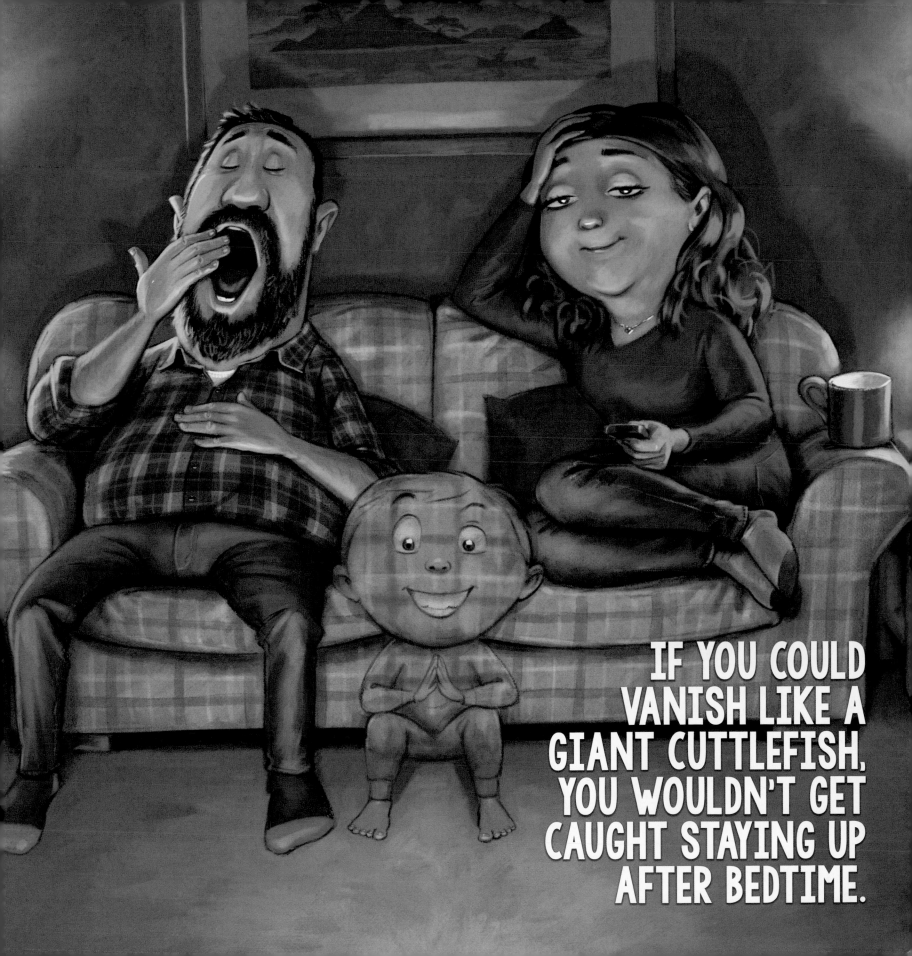

IF YOU COULD VANISH LIKE A GIANT CUTTLEFISH, YOU WOULDN'T GET CAUGHT STAYING UP AFTER BEDTIME.

WHAT YOU SHOULD KNOW

eye

arm
with suckers on
lower surface

Adult Size
Males may be larger than females, with the largest measuring about 3.2 feet including extended arms; they weigh around 28 pounds.

Life Span
One to two years

Diet
Mainly fish and crustaceans (such as crabs)

GROWING UP

A baby giant cuttlefish is called a hatchling. At two months old, it becomes a juvenile, and starting around seven months, it becomes a subadult. After mating season—May to August—adult males die. Females deposit their eggs one at a time, tucking them into rock crevices for safekeeping before they also die. The babies develop inside their eggs for three to five months before hatching. Although they are tiny, even as hatchlings, they are exact copies of their parents. Those that survive being caught by predators such as sharks grow bigger as they mature into adults.

WONDER WHY?
a giant cuttlefish sometimes has bumpy skin

A giant cuttlefish's skin has papillae (puh-pi-lee), muscle-controlled areas that can bump up, similar to human goosebumps—only the giant cuttlefish can control when, where, and for how long its skin is bumpy. Being bumpy helps a giant cuttlefish blend in when it's hiding among something lumpy or rough, such as seaweed or rocks.

mantle
Expands to pull water into its body and contracts to squeeze it out the siphon.

SUPERCHARGED!

If you could vanish like a giant cuttlefish, you'd never get caught playing hide-and-seek.

WHAT IF YOU COULD AMBUSH LIKE A LEOPARD SEA CUCUMBER?

WHERE IN THE WORLD?

The leopard sea cucumber lives around coral reefs in the Indian Ocean and the Pacific Ocean.

A leopard sea cucumber has a built-in defense system. If attacked by a predator, such as a crab, it shoots tubules from the end of its digestive track. Though its body contains several hundred tubules, it only launches about fifteen at a time. Next, its breathing system pumps these tubules full of water, so they stretch out nearly twenty times longer. Plus, on contact, cells on each tubule's surface break open and pour out a gluey substance. ZAP! The attacker is trapped. Meanwhile, the leopard sea cucumber breaks free of its tubules and escapes. Later, the lost tubules regrow, so the leopard sea cucumber is always armed.

IF YOU COULD AMBUSH LIKE A LEOPARD SEA CUCUMBER, YOU'D HAVE NO TROUBLE KEEPING UNWANTED VISITORS OUT OF YOUR ROOM.

Adult Size
May be as long as 19 inches and may weigh about 1 pound

Life Span
Up to 10 years

Diet
Mainly microscopic algae and animal wastes on the seafloor

feeding tentacles
These are inside the mouth to stay safe while moving and are pushed out to feed. They sweep in food along with sand or rock bits that pass out its digestive track with any waste.

tube feet

GROWING UP

A baby leopard sea cucumber is first called a larva. When it's older but not yet an adult, it's called a juvenile. Whenever male and female leopard sea cucumbers meet in the ocean, they release sperm (male) and eggs (female). Any egg and sperm joining produce a tiny, free-swimming larva that looks nothing like a sea cucumber. The larva spends a few weeks swimming and feeding on tiny algae. Then it settles to the seafloor and becomes a barrel-shaped juvenile that looks like a little sea cucumber. For the next few months, it feeds on more tiny algae and slowly becomes an adult leopard sea cucumber, which then grows bigger and bigger.

Similar species example provided due to lack of featured animal image.

WONDER WHY?
a leopard sea cucumber has tube feet?

A leopard sea cucumber has several rows of tube feet on its lower surface. It uses these to crawl along the seafloor. When it's climbing over something, such as a rock, the tips of its tube feet give off a little bit of a gluey substance—perfect for a firm grip.

Similar species example provided due to lack of featured animal image.

opening at end of digestive track.

SUPERCHARGED!

If you could ambush like a leopard sea cucumber, you'd be the number one tackler on your football team.

HOME SWEET HABITAT

All living things need a habitat to call home. This special place supplies the oxygen, water, food, shelter, and space needed to live and produce babies. Earth offers a number of very different habitats. What may seem like weird superpowers to us are most often ways that animals have adapted to thrive in their habitats.

- Ocean
- Coral Reef
- Freshwater
- Grasslands
- Rainforests
- Alpine

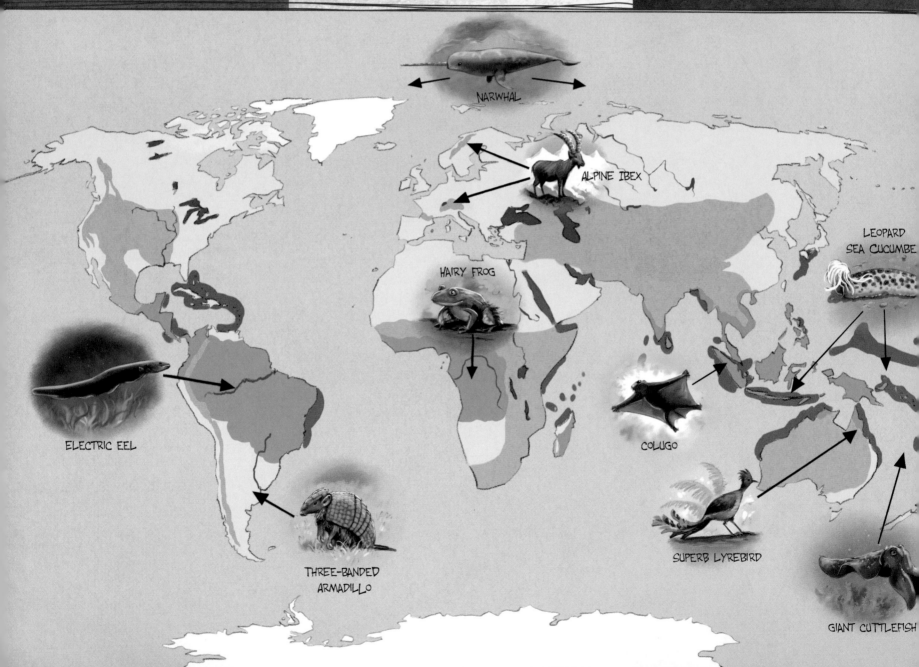

NARWHAL

ALPINE IBEX

HAIRY FROG

LEOPARD SEA CUCUMBER

ELECTRIC EEL

COLUGO

THREE-BANDED ARMADILLO

SUPERB LYREBIRD

GIANT CUTTLEFISH